Philosophy of the Biotechnological Sciences

Transhumanism and Transhuman Post Humanism

Evolution or Dehumanization?

(A critique of the Peer Review Academy process)

2nd Edition

by **Eduardo Delatorre Q.**

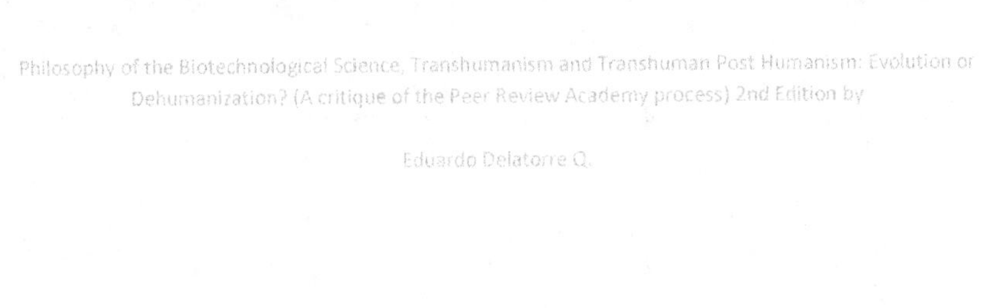

I dedicate this book to my son Edward and to my daughter Rubi Delatorre

CONTENT

(...) the one who seeks finds (...)
Gospel According to Saint Matthew chapter 7 verse 8 New American Bible

For the Lord gives wisdom;

From His mouth come knowledge and understanding.

Book of Proverbs, chapter 2, verse 6

New American Standard Bible 1995

ACKNOWLEDGEMENTS

To the entire peer review system that has shaped journal publications into what they are today: a pay-per-publish academy

BEFORE READING

This book is a commentary and critique on how slow and restrictive the academic peer review system can be. We often expect peer reviews to be thoughtful and rigorous; however, all too often, the feedback from Academic Reviewers barely rises above the level of a troll's comment on a social network.

The critique begins with the multiple rejections I received from several academic journals when trying to publish an article on the Philosophy of Biotechnological Sciences. Traditionally, Philosophy (recognized as part of the human sciences or humanities) has maintained a historical divide from the exact sciences in the last centuries —especially in relation to the Health Sciences (Medicine, Pharmacy, Biochemistry, etc.).

This book aims to bridge that gap by connecting Health Sciences with Philosophy through a revisionist critique of transhumanism.

Transhumanism is a philosophical movement that seeks to enhance the human condition—according to its advocates—through mechanisms such as genetic engineering and surgical interventions, though not limited to these methods.

The central premise I propose is: "Transhumanism, still in its beta phase, will—once it becomes mainstream—amount to nothing more than an economically exclusive modification of the human being." It will result in a "customized human" whose perceived improvements are entirely subjective.

In theory, transhumanism should strive to:

a)   Improve human health by reducing diseases.

b)   Mitigate the effects of aging that impair performance.

c)   Extend human life expectancy.

However, achieving these goals will not be accessible to everyone—particularly not to those in developing countries. Any other modifications associated with transhumanism will likewise be subjective. Even the three core goals mentioned are shaped by personal preferences—whether someone wishes to pursue them or not.

On the other hand, the transhumanist posthumanism, which aims to elevate humans to a superior state, raises a fundamental question: Is this the Übermensch Nietzsche envisioned, or something similar? If transhumanism is still in development, what can be said of a perfected version that achieves the ultimate goal of "the death of death"? (i.e. the transhumanist posthumanism)

Who will have access to this transhumanism in the near future? A person in a Latin American country living on less than three dollars a day, or only those with sufficient financial resources to afford such enhancements?

Isn't this idea of a "customized human"—tailored to individual preferences and conveniences—a form of dehumanization in itself? If achieving an optimal and ideal transhumanism is so difficult and subjective, isn't it overly utopian, idealistic, or even fanciful to believe that we could reach the Overman, whose defining trait is having conquered death?

This reflection constitutes my critique of academia and the peer review system. The unwillingness of journals to allow professionals from fields such as Humanities, Philosophy, or the Exact Sciences to address these issues demonstrates a significant limitation in the current academic landscape. What I present here defines the Philosophy of Biotechnological Sciences and leads me to conclude:

We no longer live in a Peer Review academy; we have become a Pay-to-Publish academy.

Editorial
Chevalier Champollion Publishing
PUBLICATIONS AND COMPANY

PUBLIC.0

TO THE READERS OF THIS BOOK

To all the people who have purchased this little book, I thank you and hope you find it enjoyable. Normally, under my name, I have written several academic articles as an editor or compiler of information; however, this is the first time I can directly express my way of thinking.

Hoping not to cause any further controversy, I want to make it clear that what is wonderful about the Philosophy of Biotechnological Sciences is that everything is being written as technology advances. Ignoring the reality of all the technological advancements that could affect our humanity is something we must avoid, as there is no better time than now to ask ourselves, from an academic perspective, how transhumanism and transhumanist posthumanism may affect us as individuals and as a society.

I hope you enjoy the reading, presented as a discussion or rhetoric from my perspective as the author of this text. You can see how extensive the topic is based on a comment of no more than a paragraph made by the academic reviewers, who, although not very specific or accurate, at least took the time to comment on that paper I am still trying to publish.

Eduardo Delatorre Q.

TO THE 'REVIEWERS' OF MY ACADEMIC PAPER

First, I want to make it clear that my article might be the first & ever worldwide written within the topic of: "Philosophy of Biotechnological Sciences" which is not only of a revisionist type but philosophical indeed. Note that the article will cause a debate in the critical reader about what is being a human and how transhumanism and its subsequent post humanism (in a transhumanist perspective) is nothing more than utopian, and if it was real would be defined by human subjectivity, through a customized and paid, not-Darwinian evolution. So, the understanding of what makes me human or what can make me transcend as human is linked to subjectivity more than to an universal transcendence like J. Huxley (1957) used to explain.

In order to understand my article there are some definitions that the reviewer must know (and not only knowing them but also having a clear overview about them).

What is Biotechnology?

The integration of natural sciences and engineering sciences in order to achieve the application of organisms, cells, parts thereof and molecular analogues for products and services, is defined as Biotechnology.

Source: 'biotechnology' in IUPAC Compendium of Chemical Terminology, 3rd ed. International Union of Pure and Applied Chemistry; 2006.

Online version 3.0.1, 2019.

https://doi.org/10.1351/goldbook.B00666

Comment from the author: The definition above published and that would be somehow correct, could be improved by being reduced to: "The integration of natural sciences and engineering sciences", and this is because, not all integration between natural sciences and engineering sciences might seek the "application of organisms, cells, parts thereof and molecular analogues for products and services".

So in one phrase: The integration of natural sciences and engineering sciences is Biotechnology and my article is within the biotechnological topic.

You might label my article also as "Philosophy of Biotechnology" instead of calling it as I do "Philosophy of Biotechnological Sciences"; somehow both terms have a similar approach.

Philosophy of Biotechnology as it is has been discussed in some articles that I will quote:

Henk van den Belt (2009), a Theologian, stated in a Publisher Summary: <<The term "techno science" is increasingly being used to refer to such contemporary disciplines as information and communication technology, nanotechnology, artificial intelligence and to biotechnology.

The popularity of the term is illustrated by the fact that even a group of social scientists reporting on the changing attitudes of European public towards biotechnology invokes the term to mark the departure of this particular field from the norms and values once held to be essential to the ethos of science. The evident commercialization and industrialization of biotechnology, with the pursuit of private knowledge, patents and profits, hardly meets Merton's criteria of universalism, communism, disinterestedness and skepticism, or the public's expectations about the values, accountability and social responsibility of science.

Biotechnology has become a techno science, a commercial enterprise accountable to financial markets and to shareholders. It is interesting to note that the public, according to these social scientists, have not yet fully accommodated to the new commercial realities of "techno science" and still entertain expectations about science that modern biotechnology is unlikely going to meet>>.

Source: Van den Belt, H. (2009). Philosophy of biotechnology. In Philosophy of technology and engineering sciences (pp. 1301-1340). North-Holland.

Also, Paul T. Durbin (2010), Doctor in Philosophy, and an Emeritus Professor of the University of Delaware stated:

"(…) biotechnology is not simply applied biology. It is a highly complex ensemble of relationships with genetics and biological sciences, constrain by items such as management, the state of the art at any given time, and public and political inputs. Biotechnology may be the wave of the twenty-first century, but if the twentieth century has taught us anything, scientific and technological developments are fraught with social consequences, and in a democratic society, public discussion of such issues is indeed welcome".

Source: Durbin, P. T. (2010). TOWARD A PHILOSOPHY OF BIOTECHNOLOGY: AN ESSAY. Ludus Vitalis, 18 (33).

Alfonso Barbarisi (2011), a Physician, in the Philosophy of Biotechnology (Biotechnology in Surgery), describes a "philosophy of the biotechnological approach in medicine and surgery: to consider its great value in leading and letting us be led by nature itself, by vital phenomena in order to correct its own deviances, which in medicine are diseases and phenomena in order to correct its own deviances, which in medicine are diseases and their negative consequences".

Source: Barbarisi, A. (2011). The Philosophy of Biotechnology. Biotechnology in Surgery, 1-14.

As you notice, most of the Philosophy of Biotechnology quotes in here are positive, but in what regards to artificial intelligence, transhumanism and post-humanism (transhumanist perspective), you will realize that there are some connotations that might be dangerously negative to human beings, and that is what my article is about.

Now that we have, a more opened overview about what biotechnology is we may approach to the understanding of Philosophy.

What is Philosophy? (In few words)

Quoting the Philosopher Graham Priest:

"(…) the nature of philosophy, by contrast [of the nature of breathing], is still very much an open question. One of the reasons this is so is that the nature of philosophy is itself a philosophical question, so uncontentious answers are not to be expected—if philosophers ever ceased disagreeing with one another our profession would be done for".

Source: Priest, G. (2006). What is philosophy? Philosophy, 81(2), 189-207.

Philosophy and Science

Ralph B. Winn (1942), a PhD, who is also called Ralph Bubrich, linked the Philosophy and Science, through a critic of Philosophy:

(Page 1) Many centuries ago, at the very beginning of the systematic development of philosophy, Plato declared that the thinker's domain comprises "the wholeness of things;" and indeed, the earlier thinkers took all knowledge for their province and did not hesitate to discuss problems now referred to art, psychology, economics, mathematics, or physics. Since then the meaning of philosophy has appreciably changed, however, and the intellectual descendants of the great founder of the Academy no longer claim the monopoly of all fields of study.

For there appeared in the meantime a mighty competitor-or should we say a partner?-in the pursuit of truth, namely, science. Despite the fact that the modern growth of science goes back at least to the days of Copernicus and Vesalius, science remained a junior partner of philosophy till the last decades of the 18th century; and, in recognition of its distinctive material and method, the name of "natural philosophy" was often applied to it. But with the expansion of empirical observation, measurement and experimentation, and with the development of specific scientific techniques, the junior partner revolted against the speculative character of philosophy and finally proclaimed his independence (...).

(Page 17-18)

(…) It is our opinion that the main reason for our present [1942] perplexity, for our undue modesty, for our continual retreat is that we have forgotten the lesson of Socrates. In his own day, philosophers (and sophists) devoted their attention almost entirely to the traditional problems of Substance, One and Many, Being and Becoming. Socrates revolted against this attitude and declared, in effect, that problems of philosophy are not circumscribed by learned traditions: they are problems of life, and the more vital and generally interesting they are, the more should they be studied and analyzed. The contemporary thinkers, too, deserve Socrates wrath. We are entirely too much devoted to the history of ideas and speculative fashions; we are immersed in the past and fail to notice the present. The mistake should be corrected in a true Socratic spirit. We should glance around and find concepts of vital importance to our generation of men.

Wherever there is a vital concept, there lies our work, even if it belongs nominally to the province of science.

Being this said in 1942, is not unusual to see that a reflection on sciences with a philosophical approach is not new. Hence, how can I do philosophy without questioning myself about what I see? Especially in what concerns to transhumanism and artificial intelligence.

About the submission of the article and revision process. I would like to ask to the journal Editor (or to any responsible person) of XXXX [the journal name] a few questions:

Do your reviewers have a professional background and expertise in order to understand Philosophy, Science, Biotechnology & Artificial Intelligence, but not only each discipline separately but merged as a whole concept, as an integration (Philosophy of Biotechnology or Philosophy of Biotechnological Sciences)?

As a journal (XXXX [the journal name]), how many people with a scientific approach, have you seen written articles about how artificial intelligence and transhumanism can impact [affect] in humanity?

On the other hand, I would ask my "reviewers" specifically: Have you noticed that the article that is being submitted is a 'brief communication'?

Being 'brief communication' defined as: short papers that present significant new observations of wide potential interest to readers, and that will likely stimulate further research in the field. Brief Communications may present results that are not sufficiently elaborated to justify a full Research Article but provide compelling evidence for their potential significance.

Source: SpringerLink.

https://link.springer.com/journal/44163/submission- guidelines

And now the reply per each review: All reviews are **<u>bold underlined.</u>**

To the First Reviewer

<u>The paper should be rejected, in my opinion, due to different lacks:</u>

1. Methodology:

One of the primary concerns with the paper is the lack of clarity in the methodology section. The description of the research methods is not sufficiently detailed, making it difficult for readers to understand how the study was conducted. This section should include specific information about the research design.

My first comment: Your review is not applicable, as I wrote this:

<<MATERIALS AND METHODS

This study is based on a qualitative methodological approach, focused on bibliographic compilation. No quantitative instruments such as measurement tables or surveys were used. The research was developed through a systematic and exhaustive review of the existing literature on transhumanism and posthumanism. Various academic sources were consulted, including books and digital documents. Since the research was based on secondary sources, a specific sample was not defined neither subjects were selected for the study. The bibliographical information compiled was analyzed using a critical approach, identifying recurring patterns, themes and arguments in the literature.

Special attention was paid to different perspectives on whether transhumanism and posthumanism represent evolution or dehumanization.

This methodological approach allows for a deep and critical analysis of current discussions around transhumanism and posthumanism. However, it is important to note that the findings of this study are limited to the existing literature and may not reflect all possible perspectives on these topics. It is recommended to carry out future research that adapts to scientific advances that may appear after the publication of this article>>

My second comment: I think that my method is clear enough to understand, even for an unfamiliar reader in subjects like philosophy, science, artificial intelligence, biotechnology and even with journals, papers and their texts.

2. Structure:

The overall structure of the paper lacks coherence, which affects the readability and flow of the content. Sections of the paper seem disjointed, and there is a need for a more logical progression of ideas. It is important to organize the paper in a way that guides the reader through your research in a clear and structured manner (it is the first time I saw such a"de-structured " paper). This includes providing clear section headings and ensuring that each section transitions smoothly to the next.

My first comment: Your review is not applicable, as I wrote this:

a)   I am giving an introduction of transhumanism because some people do not know about this.

The once 1300's neologism of the poet Dante Alighieri related now to the English word, "transhumanism", is possibly the first written reference to a transcendence of man to his own nature framed in a single word. Dante's phrase that would be written in English as: "Go beyond the human, it cannot be put into words", it is a precursory and quite old idea of the current transhumanism criterion.

Even I deleted the phrase "trasumanar significar per verba non sia poria", because I was told not to do any reference to a non-English language.

b) Then I do the definition of transhumanism, explaining what it is:

Transhumanism is understood as a movement aimed at the "improvement" (according to transhumanists) of man's life through a biotechnological modification of his own organism.

c) Now that people know the definition, it may has an historical background, is not a word that I invented:

In the 20th century, and already within the modern transhumanist vision, Julian Sorell Huxley, a well-known evolutionary biologist, in his 1957 work: New Bottles for New Wine4, went beyond Dante's 1300s' neologism, and mentioned the word "transhumanism" under the following context:

"We will start from new premises (...) the human species can, if it wishes, transcend itself, not only sporadically, an individual here in one way, an individual there in another way, but in its entirety, as humanity. We need a name for this new belief.Perhaps transhumanism helps: man remains man, but transcending himself, realizing new possibilities of and for his human nature."

"I believe in transhumanism": once there are enough people who can actually say that, the human species will be on the threshold of a new kind of existence, as different from ours, as ours is from that of the Peking Man. He will finally consciously fulfill his true destiny."

So, I am giving a comment of what Julian Huxley (who was an evolutionary biologist, eugenicist, and internationalist) said about transhumanism.

d) Then after quoting Huxley I give a brief critic to Huxley's point of view about transhumanism.

Please notice that Huxley wanted everyone to evolve or transcend.

But in the real world this would not be applicable, so I said:

<<Similar in some ways to Dante Alighieri's idea, but more detailed (idealistic even), Huxley sees a global human transcendence beyond an individualistic one.

Being difficult to determine what would be the limitation that we as humans could overcome (as a community or population); J. Huxley's idealism is nothing more than utopian. What limits me as a human being, perhaps another individual does not see it as a limitation, and therefore it is difficult to generalize, leaving everything framed in subjectivity>>.

23

e) If Philosophy makes people questioning a phenomenon or reality, and this is an article of "Philosophy of Biotechnological Sciences" (or call it if you want to "Philosophy of Biotechnology") after giving my opinion I want my readers to have their own criteria based on these questions that I am doing:

<<How attainable would it be for the average citizen to access an "improvement" in his (her) nature?

Would we, those who earn what is necessary to meet our basic needs (food, clothing, housing, education, etc.), be within a group economically capable of being technologically improved?

Who would define this "improvement"? We [the people] (democratically) or the pharmaceutical or biotechnological companies that [might or will] design transhumanist devices?

Would an individual citizen who can barely survive until 2024 with no more than $3 (American dollars) a day, being included in this group?

Would those families (with several members) that survive with the amount of money referred to, or even less, be within this group?

Is transhumanism an economically exclusive biotechnological evolution?>>

And as these questions seems difficult to understand to my reviewers I will answer them here:

• How attainable would it be for the average citizen to access an "improvement" in his (her) nature?

Well, if we as humans do not understand what would be transcend our limitations, and if so, if this transcend is not free or low cost, then I would be hard to pay or afford, limited only to those with money to pay for it.

• Would we, those who earn what is necessary to meet our basic needs (food, clothing, housing, education, etc.), be within a group economically capable of being technologically improved?

Obviously not, this is free market. The one who can pay can access.

• Who would define this "improvement"? We [the people] (democratically) or the pharmaceutical or biotechnological companies that [might or will] design transhumanist devices?

I work for a pharmaceutical company, and like every company in the world the customers' needs define a company operation. I cannot sell what people do not want to buy.

Somehow, this answer is limited to my context. If I were from a governmental agency or a politician, then I think that I would try to seek (if it were in my hands) to improve my people's lifestyle through transhumanism if I can obtain a benefit through it. Altruism is not common in business and politics.

• Would those families (with several members) that survive with the amount of money referred to, or even less, be within this group?

Definitely not. People would try to survive day after day and would not pay attention to a transcendence that is beyond their profits.

• Is transhumanism an economically exclusive biotechnological evolution?

Yes, it is unless and obviously, some philanthropist makes a huge project where his (or her) community transcend human limitations. But here is my premise "human limitations are subjective, what is a limitation for me, would somehow represent a triviality for other people"

Quoting what I say in my article:

<< What limits me as a human being, perhaps another individual does not see it as a limitation, and therefore it is difficult to generalize, leaving everything framed in subjectivity>>.

So, I obviously did not answer my questions, I just proposed them. Remember: "(…) stimulate further research in the field"

If I would have answered my questions, then this article would not be a "Brief Communication"

f) Then after Huxley, who published his opinion on 1957, came Esfandiary (1989)

Esfandiary published a test or a quiz to see how transhumanizable a person could be.

I explained the test in my article, in a succinct (brief) way in order to make readers to understand it.

I also mentioned that I did the test, and my result was, quote: "my score indicated "rapid growth" or a great adaptability to being transhuman"

So it is logical to go from Huxley to Esfandiary, because one represents the beginning in the 20th of transhumanism (term by the way attributed to Dante Alighieri), and Esfandiary would be some sort of a promoter of this ideology or philosophy of transhumanism.

Take into account that I mention that: "Years after Huxley, Fereidoun M. Esfandiary, better known simply as Esfandiary (or FM-2030 by his fans), and who was a writer, philosopher, futurist, professor and consultant, in 1989 (that is, 32 years after J. Huxley) in his work: "Are You a Transhuman?: Monitoring and Stimulating Your Personal Rate of Growth in a Rapidly Changing World"5, posed a series of tests and questions to assess how

<<transhumanizable>> the reader of said book could be. Seeking in this "test" or "quiz" to perhaps promote the understanding of transhumanism from our current situation and individually"

27

g) Then after a brief explanation of transhumanism and why Huxley's premise is null (the human species can, if it wishes, transcend itself, not only sporadically, an individual here in one way, an individual there in another way, but in its entirety, as humanity)

I talk about post humanism.

h) First I made a definition of post humanism but in a transhumanist perspective. So I started differentiating posthumanism (through the perspective of anthropocentrism) to the posthumanism linked to transhumanism.

<< Posthumanism, a term that goes beyond transhumanism and should not be confused with classical posthumanism (anthropocentrism in the 21st century), implies going one step further in transhumanist transcendence. It is for this reason, and for better understanding, that we will call posthumanism in this article:

Transhumanist Posthumanism. Neologism perhaps, since posthumanism related to transhumanism is usually identified (unofficially or „out of academy") as Posthuman Transhumanism. But, if we understand that posthumanism (transhumanist) implies transcending biotechnological improvement and going beyond the most complete and advanced transhuman, then the terminology proposed in the title of this article would better adapt to the idea of the transhuman that goes beyond its biotechnologically improved (so to speak) conspecifics (although this improvement could be merely subjective)>>.

And in order to avoid writing the phrase "not anthropocentric post-humanism", I decided instead to refer to it as: Posthuman Transhumanism.

i) Then after defining "posthumanism" as "Post Human Transhumanism" I give a brief explanation of what is being "post human"

<<*The transhumanist posthuman is a transhuman being whose biotechnological modifications give him (or her) suprahuman (or „overhuman") characteristics. This model, perhaps futuristic, partly fanciful, is already treated and reviewed in fully established academic, philosophical and scientific journals, among which I will mention: Journal of Posthumanism (ISSN 2634-3576 (Print); ISSN 2634- 3584 (Online)) and the Journal of Posthuman Studies (Online ISSN 2471-4461; Print ISSN 2472-4513); this until 2024. Thus, this idea of a human whose modifications allow him (or her) to surpass others, even transhumans, is already within the global academic contingent (duly indexed) and is not only mentioned (as many would believe) in fiction or science fiction literature, or its derivatives (cinema, comics, television, internet, etc.). Transhumanist Posthumanism has as its ultimate goal: "the death of the Death"*>>.

In these lines I do not only define the word "post human", I also rename this within the article context as: "transhumanist posthuman".

I explain that this "transhumanist posthuman" is somehow a sort of Nietzsche's Übermensch [Overhuman].

And I also identify this Post Human 'Overhuman' as [a] model, perhaps futuristic, partly fanciful.

You who are reading this, have notice how many transhuman devices and improvements are consumed now in our society?

Yes, almost none. So, if transhumanism itself it is not widespread then it is completely logic to state that the "transhumanist posthuman" (as I call it) (you may call it posthumanism only) is a model, perhaps futuristic, partly fanciful. And, I am right with this.

Then I mention some journals that treat the theme of posthumanism. So with this I want to make it clear that it is not an invention of mine, quote:

"Journal of Posthumanism (ISSN 2634-3576 (Print); ISSN 2634-3584 (Online)) and the Journal of Posthuman Studies (Online ISSN 2471-4461; Print ISSN 2472-4513); this until 2024."

Then I drop the phrase: "Transhumanist Posthumanism has as its ultimate goal: "the death of the Death", which is completely true. There is a book with this phrase published by SPRINGER [Journal]

SOURCE: Cordeiro, J., Wood, D. (2023). The Death of Death: The Scientific Possibility of Physical Immortality and Its Moral Defense. Alemania: Springer Nature Switzerland.

So this phrase is not mine, and I am using it within the context of posthumanism.

f) Then I provide and then answer more questions. Because within the reading of this article, it is obvious that the reader is now more familiarized with transhumanism.

Quote:

<<Are transhumanism and transhumanist posthumanism the next step in evolution?

Biologically, evolution consists of changes in the heritable traits of a population of organisms as successive generations replace each other. It is populations of organisms that evolve not individual organisms.

Under this premise, what does transhumanism imply in itself? Let us visualize it with an example: I, being a writer of scientific articles, can (and must, in order to develop a theoretical framework) read one book of 100 pages in a maximum of 2 days, and with a retentive capacity of 80% (leaving aside most of my everyday tasks).

If I could access a technology that would allow me to increase my reading capacity to a 100-page book in 10 hours (with a retention capacity of 90%), I would already be transcending my own limitations.

Obviously, it would not break the reading ability of Anne Jones, who in 2015 could read a 278-page book in 25 minutes and 31 seconds7; but for my personal purpose it would work.

Taking into account the biological definition of the word evolution, this increase in my reading ability could not be considered as "a change in a hereditary trait…" (At least not phenotypically), nor would it imply that "a population of organism evolves" but rather which would be something opposite: "an individual organism evolves" (in the example obviously the evolution would be limited to my particular and specific interests).

The case presented would also not exceed the reading capacity of Anne Jones; however, we would be before a kind of "customized (personalized) evolution".

A customized evolution, so to speak, (a neologism, this one, innovative) in which what makes me better (theoretically) than the average population depends on my subjectivity, my purchasing power, the skill or expertise of the doctor who treats me and has to perform the surgical improvement on me, the quality of the device or implant, and the skill or ability of the person responsible for the design of said device, in a special way, to be able to solve at some point any situation that requires a correction or improvement.

In the real world, Julian Huxley's premise: The human species can, if it wishes, transcend itself, not just sporadically, an individual here one- way, an individual there another way, but in its entirety, as humanity (…) would be a null premise.

In the proposed example, my increase in reading ability (being someone who is dedicated to writing) would not represent interest, for example, for someone, who works as a triathlete. It could be useful if this person needs to learn new techniques and disciplines, sports regulations, etc. But in the working life, it would not imply any significance.

Generalizing on the topic of transhumanism and posthumanism would be unnecessary>>.

About this question I will explain it, in order to you (article reviewer) could understand it better.

First, I give a definition of evolution and then I explain through an example what is, or what would be a "customized human evolution".

A customized human evolution would be that I can afford to an improvement treatment or modification in my physiology. But, this so called "improvement" would depend only in my subjectivity. An average book reader [that work with books] can surpass (in a hypothetical scenario) the fastest reader in the world Anne Jones (at 2015), but this improvement could represent nothing to a triathlete that does not have reading as his (or her) priority.

Then after this comparison, I make three additional questions answering only the last of them. But here, I will answer all, as follows:

How could the triathlete improve a physical condition without engaging in a kind of doping?

34

In a futuristic scenario, any improvement in a performance if not properly regulated would be considered some sort of doping, or use of illegal technology for performance improvement.

How legal and permissible would it be for the triathlete to access transcendence in physical performance?

This will depend on the laws applicable to the possible scenario. But this seems far from now.

Then I return to my perspective about, me, being a journal writer, wouldn't I be cheating my colleagues or competitors by improving my reading skills. And I questioned giving the answer as follows:

<<*Wouldn't I be doing the same thing, since I would cheat in comparison with other academic article writers? I answer this last question: I do not have co-workers who can feel minimized, so to speak, if my reading ability increases from one book of 100 pages read in 2 days to one book of 100 pages read in 10 hours. Perhaps my competitors (understood as those who also write academic articles in other magazines or publishers) would be at a disadvantage, but if they could access a technology in which they could read a 400-page book in 15 minutes, with a 90% of understanding (beating the record of Anne Jones who reads 278 pages in about 30 minutes) Will I be able to complain to them for having accessed better technology than mine? Will I be able to demand that they not spend their own money and resources on technologies that are possibly more expensive than the one to which I agreed? In a free market, obviously everyone buys what they can afford*>>.

This would be summarized as I previously stated within this text (page 6) that I am sending to you:

"The one who can pay can access"

g) Based on the [reader-triathlete] example I do another question with its answer:

<<*Are transhumanism and posthumanism- transhumanist criteria humanizing or dehumanizing? Regarding the issue of humanization and dehumanization. In the 19th century, the First General Etymological Dictionary of the Spanish Language defined humanize as an (active) verb that implies "to become human. To make one affable, charitable, beneficial"*

Additional and subsequent definitions of humanizing have several interesting connotations.

Humanize, according to the Cambridge Dictionary9 it is defined as: the process of demonstrating that someone has the qualities, weaknesses, etc. that are typical of a human being, in a way that makes you more likely to feel sympathy for that person (Ex: An individual describes the purpose of the story as the humanization of an antihero).

Another definition of humanizing involves: the process of making something that is not human look like a person, or treating something that is not human as if it were a person (Ex: the humanization of pets by their owners).

The opposite or inverse could be understood as dehumanizing.

Dehumanize: Make inhuman, unfriendly, and evil. Denying that someone has the qualities, weaknesses typical of a human. For the last of the definitions of humanize; the one that said: "humanizing is the process of making something that is not human look like a person, or treating something that is not human as if it were a person" (…) we could extrapolate it to the topic of artificial intelligence (A.I.).

Let us remember that a transhumanist device or a posthumanist improvement could require the use of devices or implants that work with A.I. through the Internet of Medical Things (IoMT).

If in general, we make the A.I. to "look like a person"; as it not human but we may treat it like a [mostly] virtual human being. This humanization of A.I. could also be considered a dehumanization of man.

"The humanization of A.I is the dehumanization of mankind"

This academic article embarks on the exploration of two key concepts in the context of human evolution and technology: transhumanism/posthumanism (transhumanist) and humanization/dehumanization.

The purpose is to analyze how these phenomena, driven by biotechnological advances and artificial intelligence, can affect humanity and its future.

Here is the explanation to that question:

i) First I define what is "to humanize".

ii) Then I define what is to "dehumanize".

iii) These lines might be out of context, so if you would like to I may move them, quote:

<<Let us remember that a transhumanist device or a posthumanist improvement could require the use of devices or implants that work with A.I. through the Internet of Medical Things (IoMT).

If in general, we make the A.I. to "look like a person"; as it not human but we may treat it like a [mostly] virtual human being

"The humanization of A.I is the dehumanization of mankind" *>>*

The reason of moving those lines is that I have not talked about artificial intelligence only about transhumanism.

h)   Then after the introduction and now that the reader might understand what means transhumanism, post humanism, humanize, dehumanize; then I give some short paragraphs that will summarize and introduce the reader to the article itself, quote:

<<This academic article embarks on the exploration of two key concepts in the context of human evolution and technology: transhumanism/posthumanism (transhumanist) and humanization/dehumanization. The purpose is to analyze how these phenomena, driven by biotechnological advances and artificial intelligence, can affect humanity and its future.

The work is situated in the current theoretical framework, where biotechnology and artificial intelligence are rapidly transforming our society. The background is reviewed and contributions from previous studies are highlighted to develop a deeper understanding of these topics. And, through simple cases or examples, it is proposed to give a futuristic vision (as close to reality as possible) of current transhumanism.

The justification for this work lies in the importance of these issues for our future as a species. The objectives of the study are to provide a clear understanding of these phenomena and contribute to the current discussion. It is hypothesized that transhumanism and humanization/dehumanization, as interrelated processes, will have significant implications for society>>.

i)   Then after the introduction I provided a description of the materials and methods used (I quoted this previously).

j)   Then I provide the "Results and Discussion", quote:

<<*Bibliographically, what does it mean to be transhuman?*

According the Transhumanist Manifest the transhuman is a bio- technological organism, a transformation of the human species that continues to evolve with technology. This evolution is understood within the fields of paleontology, archaeology, evolutionary biology and anthropology. It is further studied and understood in philosophical discourse and social and cultural studies.

Awareness is raised and realized through advances in technology that generate human-computer interaction, wearable devices, and computerized communication infrastructures. It is evidenced in medical science and scientific advances that identify genetic mutations and target diseases, as well as in the research and development of genetic therapies that aim to reverse and restore cellular damage to the biological system.

At an environmental level, it is experienced in space flights by astronauts who adapt to environments beyond Earth.

At an interactive level, it is experienced in the personalized use of avatars and characters from virtual reality, augmented reality, video games and other artificial environments>>

Here I am giving a long definition of "being transhuman" according to literature.

Remember that in the "Introduction" I mentioned:

(Page 3 of my article) (…) being transhuman (understood as transhuman to a biotechnologically modified individual whose adaptations would (in theory) improve their quality of life).

In the page 3 of my article, maybe it would be better if said:

(…) being transhuman (understanding 'transhuman' as a biotechnologically modified individual whose adaptations would (in theory) improve their quality of life).

Therefore, if you consider necessary, I would change those lines.

And then as a Professional in Chemistry and Pharmacy that works with medical devices (quality control, and marketing authorization process in my country's MoH) I give my own definition of transhuman:

<<A transhuman is an individual (free from pathology or physical injury related to his/her modification) who can be modified biotechnologically (whether genetically, through nanotechnology or some other suitable means), surgically, pharmacologically (but not limited to this type of modifications); at his/her own will and with the purpose of exceeding (or improving) a physical condition (whether medical, physiological, anatomical or other) in comparison with an individual (or group of individuals) whose functionality is standard according to the biological nature of their species. The modification of the transhuman must not be subjective or personalized, but useful to more than two individuals from a diverse social environment in a similar condition to the modified human prior his/her change>>.

Age effects reducing, health improvement and longevity are the three basic (but not limited to) requirements that a human modification considered as transhumanist must comply in order to be of a transhumanist source (but somehow subjectivity is very rooted in those three requirements but it does not limit transhumanism as with other modifications).

k) Then I do a question with an answer: Is a transsexual or transgender, a transhuman?

This question and section that somehow might be deleted is correct within the article scope. The unique reason that would let me to erase it is because is directly addressed to the affirmation that "human subjectivity" would define human requirements to evolve, or transcend into transhumanism.

But please take into account that this question (Is a transsexual or transgender, a transhuman?) if not written in this article may be written by someone else.

The importance of this question is that even if men (genetically born male/men) could someday have without the help of females get pregnant, this would not be transhumanism, because it does not comply any of the premises:

Of the Transhumanist Manifesto: According the Transhumanist Manifest the transhuman is a bio-technological organism, a transformation of the human species that continues to evolve with technology. This evolution is understood within the fields of paleontology, archaeology, evolutionary biology and anthropology. It is further studied and understood in philosophical discourse and social and cultural studies. Awareness is raised and realized through advances in technology that generate human-computer interaction, wearable devices, and computerized communication infrastructures. It is evidenced in medical science and scientific advances that identify genetic mutations and target diseases, as well as in the research and development of genetic therapies that aim to reverse and restore cellular damage to the biological system. At an environmental level, it is experienced in space flights by astronauts who adapt to environments beyond Earth.

At an interactive level, it is experienced in the personalized use of avatars and characters from virtual reality, augmented reality, video games and other artificial environments.

Or my definition of being transhuman: <<A transhuman is an individual (free from pathology or physical injury related to his/her modification) who can be modified biotechnologically (whether genetically, through nanotechnology or some other suitable means), surgically, pharmacologically (but not limited to this type of modifications); at his/her own will and with the purpose of exceeding (or improving) a physical condition (whether medical, physiological, anatomical or other) in comparison with an individual (or group of individuals) whose functionality is standard according to the biological nature of their species.

The modification of the transhuman must not be subjective or personalized, but useful to more than two individuals from a diverse social environment in a similar condition to the modified human prior his/her change>>.

A man that could get pregnant would be only a result of a "customized evolution", because motherhood or parenthood is part of human nature and no transcendence is implied here.

Somehow, if you want me to delete this paragraph, I would, quote:

<<The transsexual meets the criterion of undergoing a surgical modification, however, he/she would not meet the criterion of "exceeding (or improving) a physical condition (physiological and anatomical) in comparison with an individual (or group of individuals) whose functionality is standard according to the biological nature of its species."

Transsexuals or transgender individuals, no matter how much modification they undergo, and assuming that science is so advanced that it allows male transsexuals to conceive (through the union of male and female cells, or heterogamy), could not be considered transhuman from an anatomical and physiological point of view.

Subjectively, a transsexual person assigned male at birth whose modification allows them to conceive, give birth, or breastfeed a newborn would not be considered transhuman, as conception (heterogamy) is common in the human species. It is logical that not every individual assigned female at birth has motherhood as a personal objective, so no transcendence is implied for something that a female of the species would naturally do.

In a model where a phenotypically born man is biotechnologically modified (according to our definition: genetically, nanotechnologically, etc.), surgically, or pharmacologically, he might be considered transhuman if, of his own will and spontaneously, he could give rise to new life. This is because, in this case, only asexual species in nature can generate new life by partitioning or dividing themselves.

However, the premise of subjectivity could limit the definition of transhumanism. If a transsexual or transgender individual does not seek paternity (or motherhood), then procreation (even if hypothetically asexual) would not be of interest in a context of human abilities exceeding current norms. Perhaps in this specific example, we might be discussing an anatomical- physiological 'asexualization of human reproduction,' rather than transhumanism or posthumanism itself>>.

l)   The following question: <<Do transhumans exist in 2024? >> is completely correct. Here I explain the cases of Moon Rivas and Neil Harbisson, which are plenty documented. No change is applicable in here.

m)   The question: <<Beyond fiction, can a human-being get connect to artificial intelligence in a non-invasive way directly from his/her body?>> is plenty correct. No change is needed here. Neil Harbisson is connected to a device and this question would be the continuation of the so called "transhumanist" (or cyborgs, like Harbisson) and it is an introduction to the BCIs (Brain Computer Interface devices) that let people get connected to machines by EEG (electroencephalography) waves.

n)   Continuing with the last question I provide some answers. First, I introduce the BCIs into the article context. For this purpose I quote Soufineyestani, and then I mention all the things that people are currently doing by using BCIs devices.

This question and its answer are completely correct and no change is required.

o)   Then continuing with previous text I link the BCIs with a connection carried out by the Youtuber FIRESHIP in 2023.

FIRESHIP using a headset of BCIs got connected with Chat GPT by using only his EEG waves (and the headset software obviously).

I mention the process, equipment used.

p)   Then I analyze if a BCI-Chat GPT (or any other model) is a transhumanist connection, and obviously my answer is no.

Because it is not within the scope neither of the Transhumanist Manifest or within my definition:

Of the Transhumanist Manifesto: According the Transhumanist Manifest the transhuman is a bio-technological organism, a transformation of the human species that continues to evolve with technology. This evolution is understood within the fields of paleontology, archaeology, evolutionary biology and anthropology. It is further studied and understood in philosophical discourse and social and cultural studies.

Awareness is raised and realized through advances in technology that generate human-computer interaction, wearable devices, and computerized communication infrastructures. It is evidenced in medical science and scientific advances that identify genetic mutations and target diseases, as well as in the research and development of genetic therapies that aim to reverse and restore cellular damage to the biological system.

At an environmental level, it is experienced in space flights by astronauts who adapt to environments beyond Earth. At an interactive level, it is experienced in the personalized use of avatars and characters from virtual reality, augmented reality, video games and other artificial environments>>.

Or my definition of being transhuman: <<*A transhuman is an individual (free from pathology or physical injury related to his/her modification) who can be modified biotechnologically (whether genetically, through nanotechnology or some other suitable means), surgically, pharmacologically (but not limited to this type of modifications); at his/her own will and with the purpose of exceeding (or improving) a physical condition (whether medical, physiological, anatomical or other) in comparison with an individual (or group of individuals) whose functionality is standard according to the biological nature of their species.The modification of the transhuman must not be subjective or personalized, but useful to more than two individuals from a diverse social environment in a similar condition to the modified human prior his/her change*>>.

A connection Human-BCI-A.I is obviously an external device, hence no transhuman.

q)   EEG BCIs devices are not so common or widespread so I analyze another device, this one manufactured by Elon Musk, and it is the NEURALINK.

The text, approach and background of the NEURALINK and its analysis that it is not a transhumanist device is correct, so no change is required.

r)   Then I explain the ALTEREGO device, which is not invasive (no surgery required for its use). This device has been showed in many TED´s conventions, and its creator is Arnav Kapur.

The analysis, context, text, approach and background of what is mentioned about the ALTEREGO is correct and no change is needed.

Basically NEURALINK and ALTEREGO search the same, the one is non-invasive (ALTEREGO), and the other requires surgery.

Are those devices transhumanist? The answer is no.

s)   Then in order to specify "what would be a transhumanist device" I give 9 requirements that somehow must a device comply in order to be considered as a transhumanist device.

I will mention them here:

1.   Access to unlimited and validated information (keeping in mind that there is a lot of false information on the Internet), both online and in the device's memory (storage).

2.   Access to communication that allows me to have contacts with others (taking into account that the medium must be secure and have a minimum risk of being hacked or intercepted).

3.   The device should not diminish or affect my behavior (such as alienation, deterioration in my interpersonal relationships, self-harming, or harmful behavior towards others).

4.   The device should allow me to carry out economic transactions with the necessary security measures.

5.   The device must be personal and non-transferable (I cannot lend it, it cannot be stolen from me, and if it is, it could not be used by anyone other than me).

6.   The device must monitor my vital signs, and in case of emergency, it must allow me (without the need for me to be conscious) to access the necessary means to guarantee medical attention.

7.   The device should not transmit any harmful radiation.

8.   The device should not release any harmful components or particles.

9.   The performance of the device must exceed or surpass the performance of previous devices or any human and usual tasks similar to those done by said devices. A device that cannot overpass a human way of making or working could hardly be consider as a tool of improvement for a human that searches for the so called "transcendence".

But, here is a reflection (of philosophical source if you consider it):

If I have a device, despise it is advanced, the device would not make me a transhuman, because I would not fit the transhuman definition:

A transhuman is an individual (free from pathology or physical injury related to his/her modification) who can be modified biotechnologically (whether genetically, through nanotechnology or some other suitable means), surgically, pharmacologically (but not limited to this type of modifications); at his/her own will and with the purpose of exceeding (or improving) a physical condition (whether medical, physiological, anatomical or other) in comparison with an individual (or group of individuals) whose functionality is standard according to the biological nature of their species.

The modification of the transhuman must not be subjective or personalized, but useful to more than two individuals from a diverse social environment in a similar condition to the modified human prior his/her change.

But somehow if the implant or modification is plenty incorporated in the human organism (transplant recipient), then this exclusion would be subjected to further discussions.

t)   Then I compare current devices (laptops, smartphones, etc) with those like EEG-BCIs, NEURALINK and ALTEREGO, through one question.

Quote:

<<Do the previously mentioned devices or equipment (a mobile phone, a laptop, a personal computer, an EEG headset, Arnav Kapur's AlterEgo, Elon Musk's Neuralink) meet or fulfill all of the nine criteria described?

No, definitely not. Most of the performance, at least for a mobile phone, a laptop, and a personal computer, depends on human skills>>.

u)   Based on all the information compiled, as a human does this topic (transhumanism, transhumanist devices, etc.) have relevance in my life?

So, within the cultural background that I know, being South American, Latin American, and living in a development country I use through a question and an example, how long would it take to the device (that incorporates the nine statements) to be marketed in a developing country.

For that purpose I did two questions, which I think can be merged, I will write them just like they are in the article.

Quote:

<<Assuming that in a near future, there is a device that meets the 9 previous characteristics, how long would it take to reach (being commercialized in) Ecuador?

Let us suppose a utopian and quasi-futuristic scenario in which we wish to register the aforementioned device in Ecuador (the one that meet the 9 previous requirements).

Theoretically this device would be High Risk, Level IV, surgically invasive (not absorbable by the human body) and preferably long-term. The requirements for its registration in a development country (such as Ecuador) would be the following:

I. About the Manufacturing Facilities, Documents and authorizations:

a. I would need facilities with the proper operating permit issued by the ARCSA (Ecuadorian Health Care Authority) or whoever exercises its functions (this in case that the device is manufactured in Ecuador).

b. An importer or exporter license (unless of course the device is manufactured locally).

c. A certification of Good Storage, Distribution and Transportation Practices (this requirement is mandatory for those establishments that store this type of devices). If the service were outsourced, this certification would fall to the company that provides me the service.

d. A power of attorney granted by the company that manufactures the medical device, which authorizes my company (in Ecuador) to commercialize the product.

e. The plant that manufactures the device requires at least ISO 13485 certification.

f. The product must have clinical studies in the country of origin, which demonstrate that it is safe and effective in humans. If there are already three other health records issued in Ecuador for a device with similar characteristics, the clinical study would not be required.

A clinical study for new medical device would take at least 5 years (depending on the country of origin legislation).

- How long would it take for an invasive device that meets all the characteristics of the "quasi-perfect transhumanist device" that I described previously to reach Ecuador?

Being optimistic, futuristic, and somewhat realistic, perhaps by 2050 we will see something that fits what is described, arriving in a developing country such as Ecuador (Latin America) and, why not, being commercialized. By that date, many of those who may read this article in 2024 might no longer be with us>>

u) Then I return in order to close with the article to the posthuman criterion of the "Death of the Death", and I do this question with its answer:

<<If reaching a 'quasi-perfect transhumanist device' might take us 26 years from now, how feasible would it be for a transhumanized society to reach the post-human transhumanism goal of 'the death of Death'?

In a truly futuristic scenario where society has achieved disease control, life enhancement, and aging reduction, the goal of 'the death of Death' is more than utopian—it is improbable. If it were somehow feasible, it would be attainable only for those who could obtain or afford the status of having improved or transcended beyond the rest of humanity>>.

v)   Then I do a "Conclusion" of the whole article.

<<CONCLUSION

The potential that, according to transhumanists, this ideology has to expand the limits of human experience and improve the quality of life is still distant. However, it is crucial to approach these issues with caution and reflect on the ethical and moral implications of these technological advances, especially in what concerns the preservation of what makes us human.

The balance between evolution and the preservation of our humanity will depend on how society manages these changes and regulates their application.

It is essential that research and implementation of transhumanist technologies be guided by sound ethical principles and focused on the well-being of humanity as a whole, and not based on individual or fanciful requirements mostly dependent on the acquisitive power of a transhumanist consumer or client.

The human idea of emulating or surpassing deities (through the so- called 'Death of Death' implied in post-humanism), regardless of whether it is real or not, has existed since the dawn of humanity. However, the roadmap and means necessary for the transcendence of humans, being this individual more than collective, is now apparently clearly outlined>>

No change is required here; the conclusion summarizes the whole "brief communication".

So, based in this literals from a) to v), what the reviewer stated that:

2. Structure:

The overall structure of the paper lacks coherence, which affects the readability and flow of the content. Sections of the paper seem disjointed, and there is a need for a more logical progression of ideas. It is important to organize the paper in a way that guides the reader through your research in a clear and structured manner (it is the first time I saw such a "de-structured" paper). This includes providing clear section headings and ensuring that each section transitions smoothly to the next.

My comment: It is a comment lack of basis from the reviewer's 1 side.

3.   References:

The references quoted in the paper do not adequately support the research topic (all the "classics" regarding transhumanism are missing). Many of the cited works are not directly relevant to the study, which weakens the foundation of the argument and literature review.

It is essential to include recent and pertinent references that align closely with the research questions and objectives, and which provide a strong theoretical or empirical base for the study.

Comment from the author: Please kindly improve your searching methods because all the literature is available in different sources. I also provided a good bibliography, so all it is a matter for you to seek and you will find. If you want me to seek it for you and cite textually where the references were obtained let me know just in case you cannot do it by yourself.

4. Hypothesis and Aims:

The paper lacks a clear articulation of the hypothesis and aims. The paper would benefit from a clearer statement of the research problem, hypothesis, and objectives at the outset, ensuring that readers understand the purpose and significance of the work.

57

Comment from the author: This article is a brief communication. It is well structured and what you have stated is not applicable and lack of any basis.

To the Second Reviewer

<u>Reviewer: This manuscript does not meet the criteria for a scientific article and is not suitable for scientific publication.</u>

Comment from the author: This article is a "brief communication". It is well structured with a scientific approach done by a person with a professional background in medical devices (marketing authorization, sanitary registration, bioequivalence studies and quality control).

<u>The author claims that an exhaustive review of the literature on transhumanism and posthumanism has been conducted. These are two large and interdisciplinary scientific research areas. However, the manuscript only cites about 20 references, which does not make this claim very convincing. A lot more work needs to be done in terms of convincing the reader that the author is well-oriented within these disciplines.</u>

Comment from the author: This comment is not applicable, just by doing a "comment" of my article, and being brief as possible I reached almost 15 pages (just like the original article).

As a professional that works with medical devices my background allows me to have the knowledge for these disciplines. Please notice that this article is submitted as a "brief communication", so, 15 pages and 26 references are quite enough.

<u>Reviewer: The structure of the manuscript is not clear, and the paper does not read like a scientific article</u>

Comment from the author: This comment is not valid, the manuscript is clear and only small and few corrections are required, but in order to do them I need an authorization. Those corrections are stated in this document.

The article has a scientific approach within the scope of a brief communication.

<u>Reviewer: (…) (the author seems to interview themselves, posing questions to themselves that they then answer, many times without any scientific references).</u>

 Comment from the author: That "inner question" is what Philosophy of Biotechnology is about; this is not a revision article. I must give an opinion in order to let the reader discuss this work in further publications.

<u>Reviewer: (…) Lacking a coherently developed argumentation, clearly grounded in a literature review as it claims to be doing, the manuscript reads more like a speculative piece.</u>

Comment from the author: This comment is lack of validity, take a careful read at this document and you will see the whole scientific approach of the brief communication I wrote. And obviously I have to base in what other people have stated about transhumanism and post humanism, that is what Philosophy of Biotechnology is about.

Are discussions within an academic perspective forbidden in these times?

Reviewer: In sum, a lot more work needs to be done in terms of clarity, structure, convincing argument, thorough literature review, and a detailed methods section describing the study upon which the manuscript is based.

Comment from the author: This is not applicable, just a few corrections are required and by the way I am giving what to correct (move, merge) here in this pages.

Best regards,

THE AUTHOR

(Eduardo Delatorre Q.)

To the Third and Fourth Reviewers

Reviewers' comments are in *cursive*

My comments are **<u>bold and underscore</u>**

To the Editor

XXXX [Journal Name]

Ref: [XXXX]

Dear Editor

Even though the submission is over, please kindly (if possible) share my replies to my reviewer's. See the revisions in cursive and my comments bold and underscored.

To the Third Reviewer

Reviewer: Unfortunately, I cannot endorse the publishing of this paper. I recommend rejection. It does not meet the minimum standards for a publishable academic paper.

The author's comment: as a brief report and being those with these characteristics: "Brief Reports are short papers that present significant new observations of wide potential interest to readers and will likely stimulate further research in the field. They may present results that are not sufficiently elaborated to justify a full Research article, but provide compelling evidence for their potential significance", my submission seems pretty marked within the brief report scope.

Reviewer: It lacks the fundamental aspects (in terms of form, structure and substance) needed for a text to be understood as a serious academic argument.

The author's comment:

In form it is within the scope of a brief report, the structure was detailed before, first, I give a brief definition of transhumanism through history, then I provide the some sort of "official definition" and finally I provide my definition; then I mention the opinion that some people that positively identify transhumanism and I refute those opinions mentioning (being summarized) that "no transcendence is applicable for a population (as a whole) and transhumanism is subjective, individualized (customized" and is a

mean of economic segregation (only people that can pay the improvement will reach the improvement) and finally I state that transhumanism is not a type of evolution it is dehumanization implying the loss of humanity; maybe I did not referred to what make us humans, and that this a mistake that somehow would be valid because I cannot state what dehumanization is without mentioning what make us humans.

So, your revision is no applicable and needs to be improved somehow.

Reviewer: It does not provide any substantial arguments in support of the propositions made; it lacks structure -making it into a strange mix of self-interview and opinion.

The author's comment:

This comment of the reviewer is linked to the previous hence I do not need to clarify what I already had clarified. Your comment about "(…) strange mix of self-interview (SIC) and opinion" ignores the fact that this "brief report" is within the frame of a "Philosophy of Biotechnological Sciences" and how can Philosophy be done without questioning myself or the others about the fact that "transhumanism is a customized evolution, economically exclusive and that it will conduct (if physically possible) the humanity to a dehumanization and some sort of counter-evolution"

Reviewer: There is no clear question stated in the opening section

The author's comment:

I guess the questions that I propose are very clear, quote:

•   How attainable would it be for the average citizen to access an "improvement" in his (her) nature?

•   Would we, those who earn what is necessary to meet our basic needs (food, clothing, housing, education, etc.), be within a group economically capable of being technologically improved?

•   Who would define this "improvement"? We [the people] (democratically) or the pharmaceutical or biotechnological companies that [might or will] design transhumanist devices?

•   Would an individual citizen who can barely survive until 2024 with no more than $3 (American dollars) a day, being included in this group?

•   Would those families (with several members) that survive with the amount of money referred to, or even less, be within this group?

•   Is transhumanism an economically exclusive biotechnological evolution?

Reviewer: (...) and there is certainly no critical engagement with others'analysis, despite the fact that it is called (in one of its two titles...) critical analysis.

<u>The author's comment:</u>

<u>In this part I partially agree with the Reviewer #3 but some papers that are currently "academically-online" have a different approach of the one I have, and this is not bad, obviously, quote:</u>

<u>Douglas Porpora (2017) (Sociologist) that stated that "most trans-humanists are reductive physicalists", quoting More on "The Philosophy of Transhumanism" (2013).</u>

<u>And some other positive perspectives like: Max More (mentioned in the previous line) that seeks through his self invented "Extropy" that is a transhuman posthumanism with another name.</u>

Reviewer: In short, even if I approach it as a "comment" to, rather than a formulation of an independent and new argument with respect to the discourse on transhumanism and post humanism, I think some minimum requirements (of form, structure and style) should be respected.

<u>The author's comment:</u>

<u>This comment is lack of basis from the Reviewer #3 side as have been previously demonstrated in this response I am sending.</u>

Reviewer: A text lacking a purpose that can be measured in terms of scientific standards (to comment on, is not such a purpose);

<u>The author's comment:</u>

<u>Please let me know if comment on something is forbidden now. I can see an experiment and doing my commentaries which might be right or wrong. There is something that is called debate and in academic articles there is something called "discussion".</u>

<u>So once again the Reviewer #3 comment lack of basis.</u>

Reviewer: (…) that lacks a clear description of its own structure and basic divisions into seperate sections or parts, and; that lacks a coherent reference system cannot be published in an academic journal.

<u>The author's comment:</u>

<u>Idem to my previous observations.</u>

Reviewer: (..) All the more so since I find that the prior reviews of the text both have pointed this out and I see no evidence of serious revision made to the text, more than opposition and rejection of the reviewers' opinions.

The author's comment:

Seems as the reviewer thinks that this is some sort of job applicant form and only one side of the interview (or dialogue) is the one that matters.

My reviewer might be anonymous but it is obvious that he/she is biased.

Reviewer: If, as the author suggests in his response to the reviewers, this paper should not be held to the traditional standards of an academic argument - such as being transparent and evidence based - why, then, does the author pursue the path of trying to get it published as an academic journal?

The author's comment:

I think I am as transparent and scientific (or academic) enough based. Indeed this publication submission is beyond the traditional standards and arguments. And I am pursuing the path of having it published as an academic brief report because I think that knowledge and enlightenment must be share through academic means rather than social media or the news. I guess and expect that my submission is being revised by a person with the education, formation, expertise and criterion in order to give an unbiased opinion and within an academic frame.

I guess...

To the Fourth Reviewer

Reviewer: Overall, I cannot recommend this article for publication. The essay is poorly written, and the use of questions and answers is distracting and often confusing. It is not evident that the author fully understands transhumanism. I suggest formulating a clear definition of transhumanism, applying it throughout the paper, and eliminating tangential discussions.

The author's comment:

This review is lack of basis and seems pretty similar to that issued by the Reviewer #3 hence the same that I replied to the previous reviewer applies to Reviewer 4.

Here are some additional comments:

-Page 2: I do not understand how Dante's phrase is presented as a neologism in relation to transhumanism. This requires further explanation or should be omitted.

The author's comment:

I did not say that "trasumanar" is a neologism. It was at 1312. My complaint was that I even deleted that reference to the phrase "trasumanar significar per verba non si poria" because during the submission it was stated that only English must be used.

-Page 2: The section on Huxley is unclear and needs revision. Please clarify why his vision is considered utopian.

The author's comment:

Huxley considered an evolution through transhumanism in a whole population. Being this a world with high level of poverty how can all evolve through transhumanism?

Would some sort of Philanthropist offer the whole population to evolve free of charges ($)?

-Page 3: The discussion questions are unusual (e.g., "individual citizen less than 3 dollars a day"). This requires rethinking or rephrasing for clarity.

The author's comment:

Besides a possible error in grammar it is obvious that I was referring to "a person that lives with less than 3 dollars a day"

See for your reference that: Nearly a third of LAC's population lives with less than $6.85 per day (2017 PPP) according to the World Bank

(https://dataviz.worldbank.org/t/LCSPP/views/03_poverty_rate/Headcount)

There is also data that the poverty headcount ratio at Latam (Latin America or LACs) is of 3.20 U.S. dollars a day

See: https://www.statista.com/statistics/1287649/poverty-rate-latin-america/

Reviewerr: -Page 4: The analogy regarding increased reading retention and speed is weak. It is unclear how this relates to "customized evolution." Furthermore, you seem to contradict yourself, as you write, "it would not imply any significance."

The author's comment:

This was only an example in order to see that how I being an academic or professional writer within a scenario where transhumanism is physically feasible would not be interested in increasing my sports performance while this would be of interest for a triathlete.

The phrase "it would not imply any significance" that I used meant that a reading ability increasing would not be of interest for the triathlete, at least for sports, and a triathlete (being professional) might obtain earnings from the competitions. So that was the reason f the phrase:

<<In the proposed example, my increase in reading ability (being someone who is dedicated to writing) would not represent interest, for example, for someone, who works as a triathlete. It could be useful if this person needs to learn new techniques and disciplines,

sports regulations, etc. But in the working life, it would not imply any significance>>

So once again the comment of the reviewer is lack of basis.

Reviewer: -The entire introduction (the first six pages) needs to be rewritten to be more concise and coherent.

The author's comment:

Yes, it is long introduction for a brief report.

Reviewer: The brief history you provide neglects significant figures in the study of transhumanism, such as Fedorov, who is widely recognized as the father of modern transhumanism.

The author's comment:

Fedorov was a Philosopher that believed in the "Death of the Death" premise (even though not mentioned as it) and this premise is null, fanciful, and imaginative such as the "wholesome Huxley's evolution". I do not say it is impossible, I am saying that now with a transhumanism that cannot be physically, biologically or in the praxis implemented, the Death of the Death is something unreal.

But as you suggest I will mention Fedorov in subsequent submission of this brief report.

Reviewer: As a result, the historical overview is both confusing and overly dense.

The author's comment:

A lack of basis comment of the Reviewer.

Reviewer: The introduction should clearly present the methodology and thesis of the paper.

The author's comment:

A lack of basis comment of the Reviewer.

Reviewer: Additionally, there are numerous instances of awkward phrasing. For example, you write, "bibliographically, what does it mean to be transhuman?"

The author's comment:

Yes, it is okay for me to use the word "bibliographically", because in the praxis (in the real life within the current circumstances) the transhumanism is only an idea with some and interesting practices and experiments in the Genetic field specially but nothing else.

Reviewer: I suggest restructuring the introduction to include a clear overview of your research, methods, and objectives.

The author's comment:

<u>Yes, I will try to rearrange my submission as a brief report.</u>

Reviewer: Clarify whether this is a meta-study, as it is not currently clear.

The author's comment:

I do not understand what are you saying with this, please be more specific and detailed.

Reviewer: -Your use of subtitles is confusing.

<u>The author's comment:</u>

<u>A lack of basis comment of the Reviewer.</u>

Reviewer: You tend to quote seemingly random researchers without sufficient critical analysis (see pages 6-7).

<u>The author's comment:</u>

<u>If those researchers are quoted it is because from the theoretical frame I have to demonstrate that the current advances of sciences that does not allow people to be considered as transhumans.</u>

<u>So this comment from the reviewer is lack of basis.</u>

Reviewer: When you raise important questions, such as the connection between asexualization of human reproduction and transhumanism, you provide no further analysis. Moreover, there are contradictions in your argument. For instance, you state that Harbisson could be considered a transhuman, yet you also claim that he does not identify as such. This topic requires more critical engagement. The inclusion of a table for analysis is also problematic, as it is unclear why your questions are not arbitrary.

<u>The author's comment:</u>

<u>Seems as the Reviewer did not understand the context of the submission but I appreciate for the comments issued that my submission was indeed read. Read but not understood.</u>

<u>But I will comment an explanation in order for the reviewer to understand.</u>

The asexualization of human reproduction if somehow feasible or reachable is not part of transhumanism. Some people may want to procreate others not. Some of those who want to procreate might can and others not. Procreation is inherent to the human species so no transcendence is implied whether the reproduction is through sexual reproduction or through asexual reproduction even though as it were artificial.

About Harbisson, some authors consider he as transhuman, I do not consider him as a transhuman. Because there is not transcendence in through a device identify the frequency of the colors within the visible spectrum.

The table of analysis summarizes my proposal but if the Editor would have considered it as unnecessary then a "please if possible delete this table (chart)" might have been enough.

Reviewer:

-Finally, the conclusion introduces new information, which is not appropriate. Ethical and moral implications of technological advancements should be discussed in the main body of the paper rather than introduced in the conclusion. A more substantial analysis of these implications, engaging with the vibrant discussions in philosophy, theology, and science, would significantly strengthen the paper. Authors such as the controversial transhumanist Ray Kurzweil should be included in this analysis. Overall, the paper lacks clarity and contribution to the current discourse on the topic. It would also be beneficial for the author to clearly state from the outset what their paper contributes that is new.

Comment of the author:

Okay let us see the conclusion:

<<The potential that, according to transhumanists, this ideology has to expand the limits of human experience and improve the quality of life is still distant. However, it is crucial to approach these issues with caution and reflect on the ethical and moral implications of these technological advances, especially in what concerns the preservation of what makes us human. The balance between evolution and the preservation of our humanity will depend on how society manages these changes and regulates their application. It is essential that research and implementation of transhumanist technologies be guided by sound ethical principles and focused on the well-being of humanity as a whole, and not based on individual or fanciful requirements mostly dependent on the acquisitive power of a

transhumanist consumer or client.

The human idea of emulating or surpassing deities (through the so-called 'Death of Death' implied in post-humanism), regardless of whether it is real or not, has existed since the dawn of humanity. However, the roadmap and means necessary for the transcendence of humans, being this individual more than collective, is now apparently clearly outlined>>

So based on that conclusion I think that your review, quote: <<the conclusion introduces new information, which is not appropriate>> is lack of basis.

Also, what you (Reviewer #4) stated that: <<Ethical and moral implications of technological advancements should be discussed in the main body of the paper rather than introduced in the conclusion>> is not valid.

The comment by the reviewer: <<A more substantial analysis of these implications, engaging with the vibrant discussions in philosophy, theology, and science, would significantly strengthen the paper>>.

It seems a pretty good comment but I will remark the fact that my submission was a "Brief report" and it must be as succinct as possible. About theology I do not consider it necessary to mention that because I am not referring to the existence or non-existence of a divine being, I am only stating that transhumanism is something that will exclude the poor, enhance the rich and dehumanize the people. And about science, well I explained some of the current advances in Science but it is for me of interest that being this journal of Artificial Intelligence nobody has objected me the reference to Mahsa Soufineyestani, Fireship (YouTube user) and his human-A.I. connection via an EEG headset, Arnav Kapur's AlterEgo device and Elon Musk's neuralink.

Somehow, thanks for the experience and the reviews, at least you all guys took some time to check my opinion.

Best regards,

Mr. Eduardo Delatorre

Chemist, Pharmacist

SCRIPT FOR USING IN THE SLIDES OF THE SALON DANS LA VILLE
(MONTREAL BOOK FAIR, 2024)

Transhumanism is a philosophy that proposes that human improvement can be achieved through three main pathways (though not limited to these):

1) A fusion of the human body and technology (through medical devices, implants, or other suitable means).

2) Genetic modifications.

3) Medicine (among other methods) that legally enhances human performance.

///

Through the primary definition of "transhumanism," I can derive specific concepts that allow me to better understand this term.

<<Transhumanism is a Philosophy>>

What is Philosophy?

Philosophers might argue that defining philosophy in a single concept contradicts the essence of the word itself. Etymologically, philosophy means "love of wisdom," but to me, it represents a deep and structured understanding of certain phenomena, developed (or interpreted) through individuals' minds or actions within a social context.

///

According to transhumanists:

What are we, as humans?

A specie in need of improvement.

Where are we, as a species, now?

Through technological advances, we are approaching Aldous Huxley's vision of improvement (1957):

"We will start from new premises (…) the human species can, if it wishes, transcend itself, not only sporadically, an individual here in one way, an individual there in another way, but in its entirety, as humanity. We need a name for this new belief. Perhaps transhumanism helps: man remains man, but transcending himself, realizing new possibilities of and for his human nature."

///

Where will we be in the future?

According to transhumanists, the human species will achieve improvement through a fusion with technology. In some ways, posthumanists (whom I refer to as Transhuman Posthumanists) go even further with this idea (we'll explore why).

///

Is transhumanism an ideology or a philosophy?

It might be both. As an ideology, transhumanism advocates for a fusion between humans and technology. As a philosophy, it holds various deeper connotations.

///

Are there connections or disconnections between transhumanism and other disciplines?

- Theology

As a more practical field, transhumanism is not directly connected to theology. Posthumanism, however, pursues the concept of the "death of death."

- Biology

There is a tangible link between transhumanism and biology. In contrast, linking posthumanism with biology often leans more toward fiction than scientific reality.

- Artificial Intelligence

Artificial intelligence (AI) can be a component of transhumanist devices, but it is not essential (we will explore this later). For posthumanism, however, a humanized machine (or AI) or a human mind or body fused with AI—exceeding typical human capabilities—is one of its central tenets.

///

About Posthumanism

- I sometimes refer to this concept as "Posthumanism," but I believe the term "Transhuman Posthumanism" is more accurate. This helps avoid confusion with humanism in the 21st century, especially from an anthropological perspective, which views posthumanism differently.

- The aspiration to overcome death—a concept with historical roots—is one of the main pillars of Posthumanism.

- Achieving a complete fusion between AI and humans (whether mind or body) might be possible through the concept of the "singularity," where a person's mind could be migrated to a machine. While this is not achievable today, it may become feasible in the future, especially in light of advances such as Dr. Zijao Chen's work at Singapore University (MinD Vis), which uses magnetic resonance technology to convert brainwaves into images.

(For further details, see: MinD Vis online.)

- On the other hand, the idea of overcoming death remains, for now, more fantasy than science.

///

One phrase of Interest (of my book) [The First Edition of This Book]:

"The humanization of A.I. is the dehumanization of mankind."

///

Is Transhumanism Scientifically Possible?

In my book (without attributing these ideas to myself), I outline four main pillars of transhumanism:

1) Reducing the effects of aging.

2) Increasing life expectancy.

3) Overcoming diseases.

4) Enhancing average human capabilities and skills.

///

These "rules" are not mandatory, as they are tied to individual subjectivity: what I want may not align with others' desires.

I also offer a definition of what it means to be "transhuman," showing that, as of now (November 2024), there are no publicly known transhumans—and even none posthumans—in the world.

///

Definition of a Transhuman

A transhuman is an individual—free from pathology or physical injury related to their modification—who has been biotechnologically modified (whether genetically, via nanotechnology, or other suitable means), surgically altered, or pharmacologically enhanced (though not limited to these modification types). This modification is done at their will, with the goal of exceeding or improving a physical condition (medical, physiological, anatomical, etc.) compared to the standard functionality of individuals in their species. The transhuman's modification must serve more than personal or subjective benefit and be useful to at least two individuals from different social backgrounds that were in a similar condition prior to their enhancement.

///

Again: Is Transhumanism Scientifically Possible?

a) The Path of "Implementing a So-called Transhuman Device"

a.1 The subjectivity of the "pre-transhuman" (whether patient or client) will shape the specific requirements for the device design.

- For example, if someone enjoys sports more than reading, they might prioritize performance enhancements in sports over reading skills. Preferences will vary by individual.

a.2 The transhuman device must be commercially attractive (although it could also receive funding from a patron or philanthropist) (but let's be real).

- Before designing the device, it should demonstrate profit potential to justify development.

///

A Definition of Interest: What is a Transhuman Device?

A device aligned with my definition of "being transhuman."

Exclusions for a Transhuman Device

Any device or implant intended solely to correct a pathology, syndrome, or disease does not qualify as a transhuman device unless it enables the patient to exceed average human abilities compared to others in their species. Therefore, devices primarily focused on treating pathology are not transhuman.

///

This is the foundational premise before pursuing a "transhuman device."

Funding, the target demographic (who will use it), and the expectation that the device must enable users to surpass an average condition are all critical factors.

///

Once this fundamental premise is met, additional issues arise:

a.3 Timeline for Device Registration

Imagine a device with simple, established components that are already used in other devices and pose no risk or incompatibility with the human body:

- Product Design and Prototype Development: 1-5 years (depending on the device, materials, and group of scientific designers).

- Preclinical Testing: approximately 2-3 years

- Clinical Testing: 3-5 years

- Regulatory Submission and Approval (e.g., FDA in the U.S.): 1-3 years, depending on region and device classification

- Patent Issuance (U.S.): 5 years.

Total Estimated Time for Registration: 19–21 years approximately.

///

a.4 Time to Registration Post-Patent Expiration

Once the device's patent expires and other U.S. manufacturers begin production, how long would it take for these manufacturers to obtain registration?

- In a challenging scenario: approximately 4 years.

///

a.5 Selling the Device in Ecuador

If a new company (other than the inventor) wishes to sell this device in Ecuador, the device must undergo clinical studies at an institution certified by a High Surveillance Agency.

Total Estimated Time to Sell the Device in Ecuador: 2-5 years.

Overall Estimated Time for a "Transhuman Device" to go from U.S. invention to Ecuadorian market:

Approximately 25-30 years.

///

Once again: Is Transhumanism Scientifically Possible?

b) The Path of Genetic Modification (Potentially Using CRISPR-Cas9)

Currently, CRISPR-Cas9 technology is prohibited for use in germline cells (sperm or egg cells) due to ethical and safety concerns. However, in a controversial experiment, Chinese scientist He Jiankui used CRISPR-Cas9 to modify the genes of embryos via in vitro fertilization (IVF), resulting in the birth of twin girls (Lulu and Nana) and a third child. The goal was to make them resistant to HIV. This experiment was widely condemned by the global scientific community for ethical breaches and safety risks, particularly because it involved germline editing, with potential unintended effects for future generations.

///

It's worth noting that HIV transmission from parent to child can already be prevented if the father is HIV-positive, through methods such as sperm washing during IVF (in vitro fertilization) or maintaining an undetectable viral load via antiretroviral therapy (ART). These approaches prevent HIV transmission without needing CRISPR-Cas9 technology.

///

The Key Challenge of the Genetic Pathway

Suppose a genetic modification (possibly through CRISPR-Cas9) were developed, discovered, or approved that could enhance human capabilities across the American continent without targeting any specific disease or pathology.

///

Estimated Timeline for Development and Approval:

- Initial Development: 3-5 years for research and conceptualizing genetic modifications for population-wide enhancements.

- Preclinical Testing: 3-5 years for safety and efficacy tests on cells and animals.

- Clinical Trials: 8-10 years

  - Phase I: 2 years (safety in humans)

  - Phase II: 2-3 years (effectiveness)

  - Phase III: 4-5 years (long-term impact)

- Regulatory Review: 3-5 years for thorough assessment by regulatory agencies, covering safety, ethical, and societal impacts.

- Post-Approval Monitoring: 10-20 years for long-term studies on genetic stability, societal effects, and generational impact.

Total Estimated Timeline: 40 to 53 years

///

NOTES

•	A Similar Scenario with would be the registration of a "Transhumanist Medicine".

•	The main challenge is determining what specific modification or enhancement would allow humans to "transcend." This is one of the most difficult scientific barriers in applying transhumanism practically.

/ / /

Conclusion

Transhumanism could be scientifically possible if we (the scientists) overcome certain barriers:

1. What's most beneficial for the population?

2. What's most profitable for the manufacturing company?

   Would a patron, philanthropist, NGO, or foundation be willing to fund such development for the greater good?

3. The device or modification should not target a pathology, syndrome, disease, or accident injury unless it allows people to exceed normal human capabilities after implementation hence being considered as "transhumanist device".

/ / /

Cost Estimates (as a Chemist and Pharmacist)

Based on my understanding of development costs, the hypothetical expenses would be:

- Transhuman Device Development:

  - Total Development Cost: $200 million - $850 million

  - Price per Unit: $10,000 - $100,000

- Genetic Treatment Development:

  - Total Development Cost: $300 million - $1.4 billion

  - Price per Treatment: $50,000 - $500,000

EPILOGUE

The book: Philosophy of Biotechnological Sciences provides a structured reflection on transhumanism, integrating perspectives from philosophy, ethics, science, and biotechnology. The initial question of whether transhumanism is a philosophy or an ideology is an important starting point. From a philosophical standpoint, transhumanism seeks to answer fundamental questions about human nature, the meaning of enhancement, and transcendence—core inquiries of philosophy. However, as an ideology, transhumanism becomes a system of values advocating for the use of technology to improve humanity, promoting a project of transformation and progress.

The ambiguity in this duality is philosophically significant as it invites us to consider whether transhumanism is merely a set of pro-technology values (an ideology) or a genuine search for understanding of human nature and limitations (a philosophy). The answer depends on whether proponents and critics of transhumanism are truly seeking to comprehend humanity or to transform it.

The author's definition of transhumanism emphasizes improvement, longevity, and the overcoming of human limitations. This perspective is typically anthropocentric, as it places humans at the center of biotechnological goals. Such anthropocentrism has faced criticism in philosophy for limiting ethical and epistemic concerns to human interests, potentially disregarding broader impacts on the environment and other life forms. Critically, transhumanism could be seen as an extension of the modern tendency to prioritize human values over nature, promoting a divide between "human" and "natural." This invites reflection on whether transhumanism, by seeking to improve humanity, is reinforcing a model of domination over nature that has already been questioned by contemporary philosophy, particularly in ecological and posthumanist discourses.

The book also draws a critical distinction between devices and treatments that correct pathologies and those that enable users to exceed average human capabilities. This differentiation is central to transhumanism as it focuses on enhancement rather than mere treatment. The ethical dimension of this distinction has been explored by philosophers like Habermas and Kass, who caution that altering human nature, may fundamentally shift how we perceive ourselves and our values. By setting boundaries between what is considered "normal" and "enhanced," ethical concerns about equality and justice arise. What does it mean to possess abilities "above average" in a society that values equality of opportunity?

There is also an ontological issue: if an enhancement transforms us beyond typical human limits, do we remain human in an essential sense? This raises a philosophical question about identity: what constitutes "human nature" and how do we define it?

The emphasis on subjectivity in the design of transhuman devices, which must be customizable according to individual preferences, introduces philosophical questions about autonomy and authenticity. The ability to decide what to improve or change in oneself is closely related to Kantian autonomy, where individuals act according to their values and decisions rather than external impositions. However, such subjectivity could also conflict with principles of equality and justice. If transhuman devices are customized and funded privately or philanthropically, access to these enhancements may depend on socioeconomic factors, exacerbating existing inequalities. Philosophically, this prompts the question: is it ethical to allow certain individuals to "enhance" their abilities based on their desires and resources, while others do not have the same access?

The proposed timeline (33-43 years for a complete transhuman device or treatment) reveals the practical barriers transhumanism faces in becoming a reality. From a philosophical perspective, this delay can be interpreted as a fundamental challenge to the idea of linear progress inherent in transhumanism. The transhumanist project here faces the reality of contingency and the limitations of applied science. The philosophy of science has long questioned whether progress is inevitable or linear. Thinkers like Kuhn and Feyerabend have shown how scientific paradigms do not always advance towards a specific end but instead depend on social and epistemological shifts. This delay, along with regulatory barriers, suggests that transhumanism may be more of an aspirational or ideological narrative than an imminent scientific possibility.

The text concludes pragmatically: transhumanism could become scientifically possible if certain ethical and practical obstacles are overcome. From a philosophical point of view, this conclusion reflects transhumanism as an ideal rather than a current reality. This raises the question of whether the transhumanist vision is, ultimately, a technological utopia. In this sense, transhumanism shares much with past utopian projects, projecting an idealized future vision dependent on overcoming human limitations. Philosophy, reflecting on utopias and their risks, prompts us to ask whether these ideals are destined to be realized or serve merely as aspirations that can never be fully achieved. It also suggests the possibility that, in seeking to improve human nature, transhumanism may alter fundamental aspects of our identity and ethical relationships with the world.

The book offers a philosophical reflection on transhumanism within the Philosophy of Biotechnological Sciences, exploring the tension between the ideal of enhancing humanity and the ethical, social, and scientific limits it faces. Concepts of enhancement, identity, autonomy, and progress intertwine in a debate that goes beyond applied science and touches on profound questions about the essence and destiny of humanity.

FREQUENTLY ASKED QUESTIONS (FAQ) AND MORE

1. Ethical and Epistemic Concerns:

Ethical concerns refer to moral dilemmas about the use and consequences of certain actions. In the context of transhumanism, ethical concerns question whether it is right to modify human nature or create inequalities through technological and biological enhancements, affecting equality, justice, and human dignity.

Epistemic concerns relate to knowledge—how we know, understand, and validate reality. In this case, epistemic concerns question whether advances in biotechnology or artificial intelligence offer a true understanding of human enhancement or merely reflect a partial and idealized view, limiting comprehension of the long-term implications.

2. Habermas and Kass on Self-Perception and Human Values:

Jürgen Habermas: A German philosopher known for his contributions to ethics, social theory, and political philosophy. In his book The Future of Human Nature, Habermas argues that altering human nature through genetic engineering could affect our perception of ourselves as autonomous and free beings. He posits that genetic manipulation can lead to the instrumentalization of human beings, changing our relationship with ourselves and the fundamental values that allow us to live as ethical and equal individuals.

Leon Kass: An American bioethicist and former chair of the President's Council on Bioethics in the U.S. Kass argued that genetic engineering and manipulation of human nature could alter our conception of dignity and self-respect. In his book Life, Liberty, and the Defense of Dignity, Kass warns that by intervening in human nature, we might trivialize the ethical values fundamental to human coexistence, leading to a "banalization" of respect for human dignity.

3. Ontology and Ontological Problems:

Ontology is a branch of philosophy that studies the nature of being, existence, and reality. It focuses on questions about what things exist and what their essence is.

Ontological Problems: These include issues like the nature of identity (what defines an individual), existence (what makes something "real"), and categories of being (how we classify different types of entities, such as objects, ideas, or living beings).

Mention of an Ontological Problem in the Text: The text mentions an ontological problem by questioning whether enhancing human nature beyond typical limits allows us to remain essentially human. This raises questions about identity and what constitutes "human nature," as it seeks to understand to what extent enhancements might change who we are in a fundamental sense.

4. Philosophical Autonomy and Authenticity:

Autonomy in philosophy refers to a person's ability to act according to their own decisions and values, without external influences. It is a foundational principle in ethics, as it allows individuals to self-govern based on their moral principles.

Authenticity implies being true to oneself and one's values and beliefs, beyond social pressures or external expectations. Philosophically, authenticity is associated with living in accordance with a genuine understanding of one's own identity and purpose.

5. Kantian Autonomy:

Kantian autonomy is a concept in Immanuel Kant's ethics, which holds that an autonomous person acts according to principles they set for themselves rather than following external norms or desires. For Kant, true morality stems from autonomy, as an action is only truly ethical if it is done freely and out of respect for moral duty rather than for personal interests or coercion.

6. Philosophy of Science:

Philosophy of science is the branch of philosophy that examines the foundations, methods, and implications of science. It investigates how scientific knowledge is generated, validated, and applied, questioning objectivity, the nature of scientific theories, and how changes in knowledge reflect (or do not reflect) progress in understanding reality.

7. Kuhn and Feyerabend on Scientific Paradigms:

Thomas Kuhn: A philosopher of science known for his theory of scientific paradigms, presented in The Structure of Scientific Revolutions. Kuhn argued that scientific progress is not linear but occurs through "paradigm shifts." A paradigm is an accepted set of scientific practices and concepts that, when destabilized by anomalies, leads to a crisis and a scientific revolution, establishing a new paradigm.

Paul Feyerabend: A philosopher of science who questioned the idea of a single scientific method. In his work Against Method, Feyerabend argued that there is no universal scientific method and that knowledge advances in a disorderly way, often influenced by social, cultural, and even political factors. He held that science should be pluralistic, respecting the diversity of methods and perspectives.

8. Past Utopian Projects:

Some examples of utopian projects in history include:

Plato's Utopia in The Republic, which proposes an ideal society governed by philosopher-kings acting for the sake of justice and the common good.

Thomas More's Utopia, in his book Utopia, where he describes a society on a fictional island organized around values of equality and social justice, with no private property.

Utopian Socialism of thinkers like Charles Fourier and Robert Owen, who envisioned egalitarian societies where cooperation and the elimination of private property would lead to collective welfare.

These utopian projects share an idealized vision of society in which human and social limitations have been overcome to achieve harmonious and equitable coexistence, similar to the transhumanist ideal of transcending human limitations through technology and science.

If you have read this far, we send you our greetings

Chevallier Champollion Publishing

Write to us at: EDLT PUBLICATIONS

publicationsandcompany@gmail.com

Editorial
Chevalier Champollion Publishing
PUBLICATIONS AND COMPANY

PUBLIC.0

ABOUT THE BOOK

Philosophy of Biotechnological Sciences, Transhumanism, and Transhuman Post Humanism: Evolution or Dehumanization, written by **Eduardo Delatorre**, offers a rigorous philosophical and ethical exploration of transhumanism and posthumanism in the realm of biotechnological advancements. This book examines the deep questions surrounding the human pursuit of enhancement through technology, weighing whether such advancements represent genuine progress or if they risk eroding essential aspects of human identity and dignity.

Delatorre challenges the idealized promises of transhumanism by critically examining how subjective desires and financial exclusivity influence its goals. He raises fundamental questions about whether biotechnological improvements will genuinely be accessible to all or whether they will economically marginalize the majority, creating a new dimension of inequality. His analysis also introduces "Transhumanist Posthumanism," a vision of an advanced state of human existence beyond mortality, reminiscent of Nietzsche's *Übermensch*, while questioning whether such a vision is achievable or simply a utopian aspiration.

Furthermore, the book critiques the academic peer review process, which Delatorre argues can stifle innovation and limit open discourse, especially in complex interdisciplinary topics like transhumanism. He urges readers to reconsider both the ethical responsibilities and societal impacts of merging human biology with technology, emphasizing the importance of maintaining equity and moral accountability in the face of rapid technological change.

This book is both a philosophical inquiry and a critical assessment of the social and ethical dimensions of biotechnology, inviting readers to reflect deeply on the potential futures of humanity and the true cost of pursuing human enhancement.

ABOUT THE AUTHOR

Eduardo Delatorre Q. is a Latin American Chemist and Pharmacist born in Ecuador. He has written several academic articles, a Physics book with other colleagues, and various texts on drug formulation and product development under his pseudonym (or rather, second name) Francisco De Latorre Quiñónez.

Professionally, Eduardo Delatorre Q., has worked as a specialist in Pharmaceutical Validation of equipment, processes, and controlled areas in a transnational pharmaceutical company in Ecuador.

He has also worked as a Quality Control Analyst for raw materials in a mass-consumption food factory; as a Product Development Assistant and Manufacturing Supervisor in an Ecuadorian pharmaceutical manufacturing industry.

And last but not least, as a Quality Analyst in a warehouse and importer of medicines and medical devices; and as a Regulatory Affairs Coordinator specializing in obtaining sanitary registrations for generic drugs, dietary supplements, medical devices, and medical equipment, in the same company.

In recent years, the author has developed an interest in Philosophy and enjoys reading books on Philology, Etymology, Kabbalah, Exegesis (biblical commentaries on the New Testament and Talmudic commentaries), and texts ranging from Jean-François Champollion to Karl Josias Von Bunsen in the field of Near Eastern studies.